Smail LAMINE

General ecology tutorials

Smail LAMINE

General ecology tutorials

General ecology tutorials

ScienciaScripts

Imprint

Cover image: www.ingimage.com

This book is a translation from the original published under ISBN 978-620-6-72818-4.

Publisher:
Sciencia Scripts
is a trademark of
Dodo Books Indian Ocean Ltd. and OmniScriptum S.R.L publishing group

120 High Road, East Finchley, London, N2 9ED, United Kingdom
Str. Armeneasca 28/1, office 1, Chisinau MD-2012, Republic of Moldova, Europe
Managing Directors: Ieva Konstantinova, Victoria Ursu
info@omniscriptum.com

Printed at: see last page
ISBN: 978-620-8-54289-4

Table of

Preamble

This tutorial handout for the General Ecology module is intended primarily for second-year preparatory students (Semester 3 - First Cycle - SNV) at the Ecole Supérieure d'Agronomie de Mostaganem.

Methodology Teaching Unit

- **VH:** 10:30 p.m.
- **Coefficient**: /
- **Credit: /**

Teaching aim : To apply certain basic scientific concepts, through all the proposed exercises in the subject taught, i.e. "General Ecology.

Assessment : 25% continuous assessment (TD exam+ trip report).

Recommended prior knowledge: Biology, Ecology and Zoology

Document content :

Part 1: A reminder of some basic ecological ;
Part 2: Sampling ;
Part 3: Biocenotics: application ;
Part 4: Study of cohabitation ;
Part 5: Biotic factors / intra-interspecific relationships ;
Part 6: Food chains and food webs ;
Part 7: Geographical distribution of certain rodent species according climatic conditions.

Specific objectives

On completion of this course, students will be able to :

✓ Synthesize or construct essential ideas and draw up action plans to remedy ecosystems affected by natural and/or anthropogenic stresses.

✓ Raise awareness among the entire human population of the need to respect the environment and biodiversity (mainly flora and fauna, and especially species known as key ecosystem species);

✓ Propose action plans for decision-makers to implement ecologically correct practices (e.g. sustainable development).

To evaluate this tutorial handout, we proposed ***Dr. BOUZID Khadidja*** of the Ecole Supérieure d'Agronomie de Mostaganem and ***Dr. GHARABI Dhia*** of the Université Ibn Khaldoun-Tiaret.

Introduction

General introduction

Ecology is a science of complexity, attempting a synthesis between the living and the non-living. To achieve this, it has its roots in thermodynamics and Darwinian evolution. The former identifies the primary resources and constraints that determine the framework within which organisms can survive, grow and reproduce. The latter explains the innumerable strategies by which living beings accommodate themselves to their environment, i.e. resist its negative effects and make the most of available matter, energy, space and time (Tirard et *al.*, 2016).

This General Ecology tut handout is presented in seven main parts (TDs), each sufficiently illustrated to facilitate comprehension. The first tutorial covers some fundamental ecological concepts. The second TD introduces the notion of minimum area, with a description of the main sampling techniques and methods used in the animal kingdom. The third TD deals with applied biocenotics, covering several ecological parameters of populations and stands (density, relative abundance (frequency), constancy and the different types of spatial dispersal of species).

The fourth TD is devoted the study of cohabitation between species, as well as a number of comparative ecological indices (Sørensen's index, Jaccard's index (or coefficient), Odum's index, etc.) used to establish the degree of association between two species. The fifth TD deals with the set of relationships that can exist between different species, which can be summed up as biotic factors / intra- and interspecific relationships. The sixth TD looks at the different types of food chain and their importance in transferring energy from one trophic level to another, and also in maintaining the stability of terrestrial and aquatic ecosystems.

Finally, the seventh section looks at the geographical distribution of certain rodent species in relation to climatic conditions.

This handout is made available to students as a teaching aid to facilitate tutorial sessions and shed more light on certain course concepts that are often poorly assimilated by long face-to-face and distance learning sessions .

TD n° 01: Review of a few basic ecological

Part 1	*A reminder of some basic ecological concepts*

TD n° 01: Review of a few basic ecological

Introduction

The diversity of organisms is not limited to their morphological and anatomical characteristics. Growth, age at maturity, number of offspring per reproduction, dispersal and life expectancy are all traits whose variations are part of the extraordinary diversity of living organisms (Barbault, 2008).

1. *Some definitions Individual*

It is a functional biological system that in simplest case is reduced to a single
cell. As it grows, the organism must adapt to the various conditions of the environment in which it lives.
he lives.

Species

It refers the set of individuals similar in appearance, habitat, fertile
and usually sterile with individuals of another species.

Population

A group of individuals of the same species living in a given territory at a given time, characterized by a structure, organization, functioning and evolution, which are themselves controlled by the environment.

Settlement

Individuals of several species of the same systematic group occupying a single area.
called a biotope (examples: mites, birds, mammals, etc.).

Ecology

The term "ecology" was coined in 1866 by German biologist Ernst Haeckel, from two Greek words: ***Oikos***, meaning house, habitat, and ***Logos***, meaning science. Ecology thus appears as the science of habitat, studying the conditions of existence of living beings and the interactions of all kinds that exist between these living beings and their environments.

To situate this science in relation to the other biological sciences, it is useful to consider the various levels of organization of living matter.

The simplest system with all the fundamental characteristics of living beings is the cell. Cells associated to form tissues and organs are integrated into multicellular animal or plant organisms.

At a higher level, the individuals of certain species can colonies or societies. At an even higher level is the population, for all living beings.

The populations of the various species - bacterial, plant and animal - form an even more complex level of organization, the biological community or biocenosis, which occupies the biotope and together constitute an ecosystem.

The ultimate level of organization in the living world is made up of all the ecosystems of the planet, i.e. the ecosphere or biosphere.

2. *Notion of ecosystem*

An ecological system, or ecosystem, was defined by English botanist Arthur Tansley in 1935. It's a biological system made up of two inseparable elements: the biocenosis and the biotope.

The biocenosis is the set of organisms living together (zoocenosis, phytocenosis, microbiocenosis, mycocenosis...).

The biotope is the fragment of the biosphere that provides the biocenosis with the abiotic environment essential for its establishment and development. It is also defined as the set of ecological abiotic factors (substratum, soil "edaphotope"; climate "climatope") that characterize the environment where a given biocenosis lives.

The ecosystem is the basic structural unit of the biosphere, and it is possible to analyze structure and operation.
Example of an ecosystem: forest.
Biocenosis: group of living organisms (phytocenosis (trees, shrubs, herbaceous plants, mosses, etc.); zoocenosis (animals), mycocenosis, microbiocenosis, etc.).

Biotope: all the elements, essentially abiotic, that make up the forest environment, such as edaphic factors (soil), geographical factors (latitude, altitude, slope, exposure, etc.), climatic factors (temperature, precipitation, humidity, etc.), chemical factors (pH, mineral salt content, etc.), etc.

3. *Subdivisions of ecology*

Ecological studies conventionally focus on three levels of integration: the individual, the population and the community. It is these levels and the specificities of the studies undertaken that determine the different disciplines (divisions) in ecology (Faurie et *al.*,2011).

Autoecology or factorial ecology: this is the discipline ecology that studies the relationship between a single species and its environment. It defines the tolerance limits and preferences of the species studied in relation to various ecological factors, and examines the action of the environment on morphology, physiology and ethology.

Demoecology, population ecology or population dynamics: focuses on the qualitative and quantitative characteristics of populations. It analyzes variations in abundance (demography), spatial distribution, population productivity, social relationships and relationships with the environment.

: the study of living . It analyzes the relationships between different species and their environment. It covers various ecological aspects such as predation, competition, parasitism, the evolution of biocenoses and their productivity.

Modern ecology has seen the birth of synecology, a discipline of ecology dedicated to the study of the structure and functioning of ecosystems. The integration of computer tools has led to the birth of digital ecology, which deals with the modeling and simulation of ecological systems. Modern technological approaches have led to major advances global ecology, focusing on the study of structure and function on a biospheric scale (Ramade, 2009).

Conclusion

Ecological systems are made up of interconnected populations and a background star of the physical-chemical environment. The reality of these systems is materialized by the permanence, to a greater or lesser degree, of the populations and their interrelationships. The key to this permanence, beyond ephemeral existence of individuals, lies on the one hand in the hereditary transmission of their traits, and on the other, paradoxically, in the genetic and phenotypic variability of populations, thanks to which they can adapt to changes in their physico-chemical and biotic environment.

TD n° 02 : Sampling

Part 2 *Sampling*

TD n° 2 : Sampling

Introduction

For practical reasons (time, available resources, number of field staff, etc.), it is not possible to exhaustively measure the size of a population across the entire ecosystem of which it is a part. In such cases, samples are taken corresponding to specific volumes or areas.

The sample size and method of counting individuals must obviously be adapted to the population under study. In each case, it is important to assess fidelity, precision and accuracy of the technique envisaged (Marc Mazerolle, 2019).

1. Definition

In ecology, sampling is a procedure, generally carried out in the field, which consists of collecting samples (data). A sample is a fragment of a whole (an ecosystem, for example) taken to study the whole itself (Vaillant, 2005).

2. Purpose of sampling

Sampling is generally undertaken to identify the species present, as well variations in their densities over time and space.

The ecologist must first choose a suitable sampling method before proceeding.
à :

- Identification of captured organisms ;
- Counting the number of organisms caught ;
- Data analysis. The purpose of data analysis is to:

– Knowledge of flora and fauna;
– Determine the type of distribution of individuals **(Figure 1)**;
– Estimate the total population of a given area or volume;
– Interpret quantitative and qualitative differences between samples.

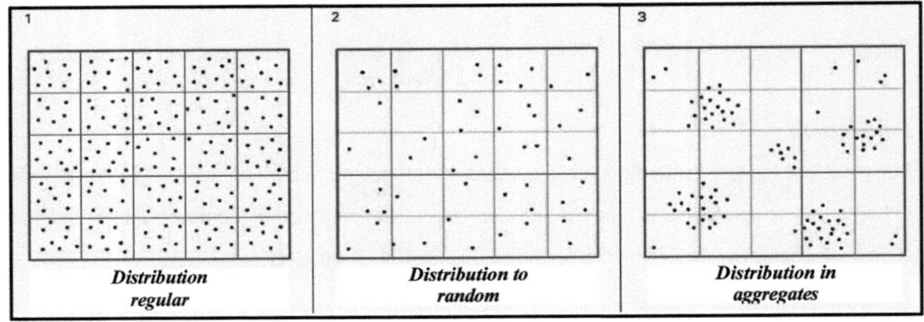

Figure. 1. Main types of spatial distribution of individuals making up a population.

3. Some sampling

3.1. Terrestrial fauna

To study lion populations on a savannah, for example, observation from a distance with a pair of binoculars will suffice (direct method).

To study a terrestrial fauna of lesser , such as rodents by

For example, the ecologist uses tipping cages containing bait.

3.2. Endogeous fauna (fauna)

A volume of soil (core sample) is taken using a soil auger. The fauna is then extracted from the soil using the Berlèse-Tullgren extractor **(Figure 2**).

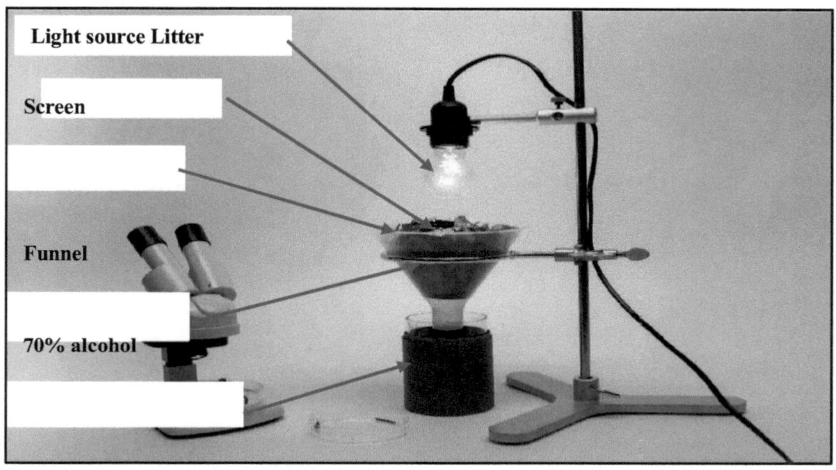

Figure 2. Schematic diagram a Berlèse-type device for capturing soil and water fauna. litter (Ramade, 2009).

3.3. Epigeous fauna (occupying the soil surface)

For arthropods living on the soil surface (arachnids, insects), for example, we can use the quadra method, the principle of which is based on collecting all individuals of all species present within a sample area (or quadra), defined beforehand and materialized by a device (tool) made of metal, wood or plastic. The sampling procedure is carried out on a homogeneous plot of land, and the position of the quadra within this area is obtained at random by discarding the device.

3.4. Aquatic fauna

Fauna (insects, crustaceans, etc.) occupying the watercourses are collected using nets.

Surber and Troubleau (**Fig. 3**). These nets have 275µm mesh.

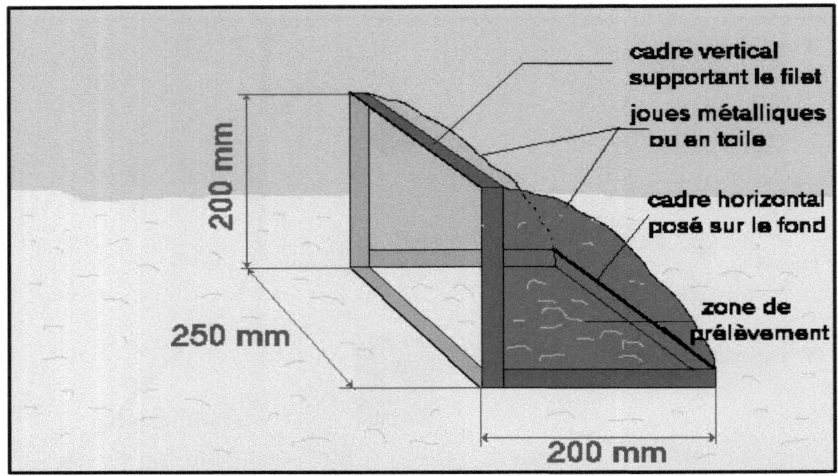

Figure 3: Diagram of Surber-type net for collecting benthic fauna (water bottom) of watercourses (Ramade, 2009).

3.5. Birds

Breeding birds in a forest, for example, can be sampled by listening to their song for 20 minutes (minimum listening time). This is usually done twice (early and late spring). The bird's song is known advance by the ornithologist, who simply ticks off the species each time he or she listens. This method is known as IPA (Indice Ponctuel d'Abondance).

4. Sampling problems

The real problem with sampling is defining a work surface that must be :

- Big enough to work in and small enough to finalize the study ;
- Homogeneous (identical physical and biological characteristics in all). Ex: the herbaceous stratum as a homogeneous surface, no trees, no rocks).

Remarks

- Sampling should not be carried out in the edges (ecotones);
- For slow-moving animals, the surface should be small and vice versa;

-For low-density species, the work surface must be large. Ex: lions, animals...

5. *Representative sample*

To obtain representative samples, we must :

- Respect sample size, which is determined the minimum sampling area;
- Repeat samples in time and space.

5.1. *Determining sample size*

- Clear the ecosystem and define a large, homogeneous area, avoiding edges;
- Start with sample 1 (first surface) and take all the species.

present ;

- Take ech. 2 (same area as ech. 1 but in a different location, still within the large homogeneous area) and take only the new species that appear (different from those in ech. 1) and add them together (new species + species in ech. 1);
- Repeat the operation several times until no more new species appear, i.e. a constant accumulation, at which point we stop sampling and the minimum area is reached.

5.2. *Definition minimum area*

This is the smallest homogeneous surface representative of a given biocenosis, in which all species without exception are present (miniature biocenosis). For the fauna of the herbaceous stratum, for example, it is $1m \times 1m = 1m^2$.

5.3. *Importance of spatial sample distribution*

Due to the heterogeneity of ecosystems, the spatial distribution of individuals in a population is rarely uniform: their density varies from one place to another.

Under these conditions, no single measurement is sufficient to ensure the representativeness of a result. This difficulty is overcome by repeating samples in space, with the aim of :

- Firstly, to identify the average characteristics of the population;
- Secondly, to determine the amplitude of variation of measured quantities.

5.4. *Importance of sample distribution over time*

We need to sample at different times of the year (4 seasons) to take into account fluctuations in density within a given population and the different physiological stages of the species studied.

6. *Application exercise*

Determine the minimum area for sampling the entomofauna of a herbaceous stratum. Thirty-two samples are taken, and table below records all the species in sample 1 and any new species that appear in the next sample at each stage. Draw the cumulative species curve as a function of sampling area.

Table 1. Application exercise for determining minimum area

Sample number	New species
Sample 1	10
Sample 2	6
Sample 3	3
Sample 4	4
Sample 5	1
Sample 6	2
Sample 7	1
Sample 8	2
Sample 9	1
Sample 10	1
Sample 11	1
Sample 12	2
Sample 13	1
Sample 14	1
Sample 15	0
Sample 16	1
Sample 17	1
Sample 18	1
Sample 19	1
Sample 20	0
Sample 21	0
Sample 22	0
Sample 23	1
Sample 24	0
Sample 25	0
Sample 26	1
Sample 27	0
Sample 28	0
Sample 29	0
Sample 30	0
Sample 31	0
Sample 32	0

Source: Lamine, 2021

7. Correction of the application exercise

Table 2. Exercise correction - minimum area

Sample number	New species	Cumulative species
1	10	**10**
2	6	**16**
3	3	**19**
4	4	**23**
5	1	**24**
6	2	**26**
7	1	**27**
8	2	**29**
9	1	**30**
10	1	**31**
11	1	**32**
12	2	**34**
13	1	**35**
14	1	**36**
15	0	**36**
16	1	**37**
17	1	**38**
18	1	**39**
19	1	**40**
20	0	**40**
21	0	**40**
22	0	**40**
23	1	**41**
24	0	**41**
25	0	**41**
26	1	**42**
27	**0**	**42**
28	0	**42**
29	0	**42**
30	0	**42**
31	0	**42**
32	0	**42**

Note: the cumulative species must be calculated by the students and then they will produce the graph in question: cumulative species as a function of sampling area.

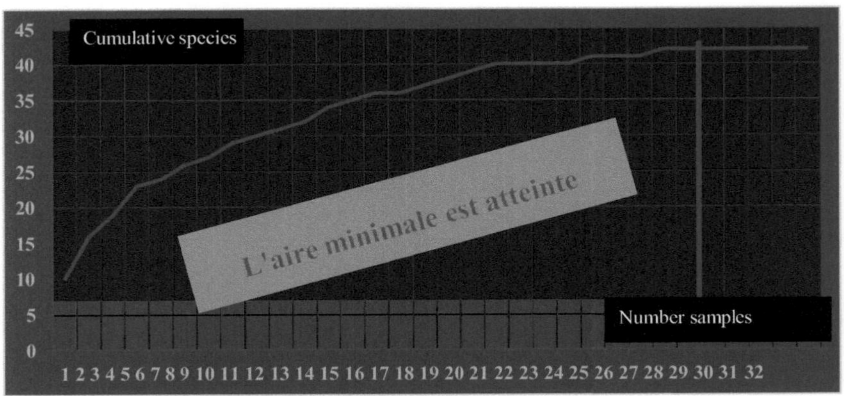

Figure 4: Minimum area for sampling the entomofauna of a particular area herbaceous stratum

The minimum area calculated therefore= 27× the area sample 1 ($1m^{2)}$.

Conclusion

A sample is a fragment of a whole, taken to assess the whole. Numerous methods of observation and measurement applied to such fragments can be proposed, adapted to each particular case with a view to obtaining a satisfactory representation of the object studied. Sampling must be adapted to test the hypothesis made, on a given spatial and temporal scale, about the structure or dynamics of the biological system under study. It is imperative to take the time to plan your sampling.

TD n° 03: Biocenotics: application

Part 3 *Biocenotics: application*

TD n° 3 : Biocenotics - application

Introduction

The distribution of individuals in a population describes how they are distributed in relation to each other in the space occupied by the population. This distribution is of considerable ecological interest, as it provides indications of the extent of intraspecific competition, dispersal ability and different reproductive systems (Tirard et *al.*, 2012).

1. A few points to remember

The population is the set of individuals of the same species living in a given territory at a given time, characterized by a structure, organization, functioning and evolution, which are themselves controlled by the environment.

The ecosystem is the whole formed by a community of living beings, known as a biocenosis, interrelated with its environment, the biotope.

Biocenotics is the discipline of ecology concerned with the study of communities of living beings.

2. Ecological parameters

Several parameters are used in ecological studies. The main ones are :

2.1. Density

The most precise measure for counting individuals of species forming a biocenosis is density. This is established by counting the number of individuals on a given surface or in a given volume. It is therefore a value reduced to a unit of surface area or volume, not to be confused with relative abundance (see below).

In the case of epigeic fauna, for example, quadras are first defined (shape and area) and individuals are counted within them. The number of individuals is then converted into density. The total number of individuals is calculated by multiplying the density by the total surface area.

2.2. Relative abundance of species (frequency)

To determine species abundance, we calculate frequency, which represents the percentage (%) of individuals of a given species in relation to the total number of individuals of all species sampled. It is calculated as follows

$$F_i = (N_i/N_t) * 100$$

F_i = frequency species i ;

N_i = number of individuals species i ;

N_t = total number individuals of all species.

Frequencies are divided into 5 classes (Dajoz, 2006). $0\% > F_i < 20\%$: rare species ;
$20\% > F_i < 40\%$: rare and scattered species; $40\% > F_i < 60\%$: species is not very abundant; $60\% > F_i < 80\%$: abundant species ;

$F_i \geq 80\%$: a very abundant species.

2.3. Consistency

This is the species' fidelity. It represents the number of surveys in which a species is present in relation to the total number of surveys carried out (Dajoz, 2006). It is given by the equation below:

$$C_i = (P_i/P) * 100$$

C_i: constancy species i ;

P_i: number of surveys where species i is present; P: total number of surveys carried out.

There are three classes of *constancy*: $C_i > 50\%$: species is constant ;
$25\% < C_i < 50\%$: accessory species; $C_i < 25\%$: accidental species.

2.4. Spatial dispersion of a population

The distribution of individuals in a population describes the distribution space of the individuals in this population.

There are 3 basic distribution methods:

- uniform (or regular) distribution ;
- grouped (or contagious, or gregarious) distribution;
- random distribution.

2.4.1. Even distribution

Uniform distribution tends to distribute the individuals in a population uniformly.

It can be achieved artificially in many crops. It can also be observed in cases of very severe intra-specific competition, where the distribution approaches a uniform distribution.

In territorial animals (lions, foxes, etc.), individuals are often evenly spaced. Sedentary invertebrates may also disperse regularly over small areas (Chironomid larvae, Trichoptera...).

Moreover, a regular distribution rarely describes the dispersal of a population over a large area. Example: Fish like sticklebacks, which defend a territory, are highly individualistic.

2.4.2. Contagious distribution (aggregated or grouped)

This is the most frequent distribution. It affects many animals. It has two causes
essential :

- a heterogeneous environment, with conditions that are unevenly favorable to the species; aggregation can occur during migrations, in search of food or water, or defending a territory (as in the case of ladybugs, social insects, etc.). It can be observed in plants with no effective means of dissemination (vegetative propagation), or by fragmentation of a single plant (as in the case of strawberry plants, which multiply by stolons).

2.4.3. Random distribution

It is exceptional in nature and occurs in very homogeneous environments. In such a case, the individuals in the population have an equal chance of occupying any point in the environment. The presence of one individual therefore has no influence on the position of neighboring individuals. This assumes that environmental factors have little or no effect on the dispersal of individuals in the population, who therefore tend neither to avoid each other nor to congregate. This is often the case for the arrangement of insect eggs (or young larvae).

The spatial distribution of individuals can be assessed by comparing variance
δ^2 to the mean $\bar{X}$.

Consider an environment in which N samples have been taken from identical surfaces, each
containing X number of individuals.

If $\bar{X}$ is the arithmetic mean of the number of individuals in sampling set,
the variance of the distribution is given by :

$$\delta^2 = \sum_{1}^{N} (n_i * x_i^2)/N - \bar{X}^{(2)} \qquad \textbf{with} \qquad \bar{X} = \sum_{1}^{N} (n_i * x_i)/N$$

N: total number of samples ;

x_i: number of individuals in any sample; n_i: number of readings in any sample; $\bar{X}$: sample mean.

If :

$\bar{X} > \delta^2 \cong 0$: the distribution is uniform (regular); $\delta^2 > \bar{X}$: the distribution is aggregated (contagious);
$\delta^2 \cong \bar{X}$: the distribution is random.

3. *Application exercise*

Table 3. Biocenotic application exercise

Prel. / Esp.	1	2	3	4	5	6	7	8	9	10	11	12	13	14	15	16	17	18	19	20	21	22	23	24	25	26	27	28	29	30
A	6	1	5	3	0	0	7	6	3	2	0	1	2	3	12	0	8	10	0	18	17	10	0	9	0	0	14	17	0	9
B	18	7	2	1	0	3	4	0	5	5	0	7	10	6	10	8	0	9	14	15	17	1	0	19	0	18	17	1	2	0
C	0	0	0	0	0	3	5	0	0	9	0	1	0	0	0	8	2	7	0	0	16	15	0	0	19	17	13	0	0	0

Source: Lamine, 2013

Questions :

1. Calculate the relative abundance of the three species: N_A, N_B and N_C;
2. Calculate the frequency of species A, B and C;
3. Calculate the constancy of species A, B and C;
4. Determine the type of spatial distribution of the species studied;
5. Study the cohabitation between pairs of the following species: A-B, A-C and B-C.

4. *Correction of the biocenosis exercise*

1. Calculation relative abundance of three species: N_A, N_B and N_C

N= **NA=163**	NB= **NB=199**	NC= **NC=115**	Nt= 487
A= **A=21**	B= **B=23**	C= **C=12**	

F= N_{esp} / Nt x 100 (N_{esp}= Nt=)

F_A=N_A / Nt x 100 = 163/477 x 100= 35% ➡ Rare and scattered species (between 20% and 40%); FB=NB / Nt x 100 = 199/477 x 100= 42% ➡ Sparse species (between 40% and 60%); Fc=Nc / Nt x 100 = 115/477 x 100= 24% ➡ Rare and scattered species (between 20% and 40%).

2. Calculating constancy

C = p_{esp} / P x100 (p_{esp}= the number of surveys where the species is present, P=number of all surveys) ;

C_A= p_A / P x 100=21 / 30 x 100= 70% ➡ Constant species (C > 50%); C_B= p_B / P x 100=23 / 30 x 100= 76% ➡ Constant species (C> 50%);

C_C= p_C / P x 100=12 / 30 x 100= 40% ➡ Accessory species (25%< C< 50%).

3. Determining the spatial distribution of species

This is the comparison of the &(2) variance with the X of the sample:

X=Σ ni xi / N   **&(2)= Σ ni xi²/ N - X²**

If

- **&(2)= X ⟶ Distribution is random ;**
- **&(2)< X ⟶ Distribution is regular ;**
- **&(2) > X ⟶ Distribution is contagious.**

a) Order the series ;

b) Calculate the range of the series, n= max value - min value ;

c) Determine the number of classes in the series, K=$\sqrt{N}$;

d) Determine the step size, d> n / K.

species

a) Order the series

0 (9),1 (2),2 (2),3 (3),4 (0),5 (1),6 (2),7 (1),8 (1),9 (1),10 (2),11 (0),12 (1),13 (0),14 (1),15 (0),16 (0),17

(2),18 (1),19 (1).

b) Number of classes: K=$\sqrt{N}$= $\sqrt{30}$ = 5 ;

c) The scope

n= max value - min value= 19 - 0= 19   n=19 ;

d)

e) Pitch:

d > n / K   d > 19 / 5 d= 4.

The statistical table

Classes	xi	or	nixi	nixi²
[0-4[	2	16	32	64
[4-8[	6	4	24	144
[8-12[	10	4	40	400
[12-16[	14	2	28	392
[16-20[	18	4	72	1296
Σ		30	196	2296

X= 196/30=6.53   X=6,53

&(2)= 2296/30 - 42.68= 33.85 &(2) =33.85

&(2)> X contagious distribution.

species

a) **Order the series**

0 (7),1 (3),2 (2),3 (1),4 (1),5 (2),6 (1),7 (2),8 (1),9 (1),10 (2),11 (0),12 (0),13 (0),14 (1),15 (1),16 (0),17 (2),18 (2),19 (1).

b) **Number of classes**

K= $\sqrt{N}$=$\sqrt{30}$= 5 ;

c) **The scope**

n= max value - min value= 19 - 0= 19 n=19 ;

d) **Pitch**:

d > n / K d > 19 / 5 d= 4.

The statistical table

Classes	xi	or	nixi	nixi^2
	2	13	26	52
[4-8[	6	6	36	216
[8-12[	10	4	40	400
[12-16[	14	2	28	392
[16-20[	18	5	90	1620
Σ		30	220	2680

X= 220/30=7.33 X=7,33

&(2)= 2680/30 - $(7.33)^2$= 35.6 &(2) =35.6

&(2) > X contagious distribution

For

a) **Order the series**

0 (18),1 (1),2 (1),3 (1),4 (0),5 (1),6 (0),7 (1),8 (1),9 (1),10 (0),11 (0),12 (0),13 (1),14 (0),15 (1),16 (1),17 (1),18 (0),19 (1).

b) **Number of classes**

K= $\sqrt{N}$=$\sqrt{30}$= 5 ;

c) **The scope**

n= max value - min value= 19 - 0= 19 n=19 ;

d) **The step**

d > n / K d > 19 / 5 d= 4.

The statistical table :

Classes	xi	or	nixi	$nixi^2$
[0-4[	2	21	42	84
[4-8[	6	2	12	72
[8-12[	10	2	20	200
[12-16[	14	2	28	392
[16-20[	18	3	54	972
Σ		30	156	1720

X= 156/30= 5.2 X= 5.2

&(2)= 1720/30 - $(5,2)^2$= 30,29 &(2) =30,29

&(2) > X contagious distribution.

Conclusion

Like populations and stands, they can be broadly defined by certain ecological parameters: density, relative abundance, species richness, constancy, type of spatial distribution and biomass. Such approaches are generally developed during ecosystem studies of the type already mentioned in previous sections.

Distribution table for X^2 (K. Pearson distribution).

Table de distribution de χ^2 (loi de K. Pearson)

La table donne la probabilité α, en fonction du nombre de degrés de liberté ν, pour que χ^2 égale ou dépasse une valeur donnée χ^2_α.

$$\alpha = P(\chi^2 \geq \chi^2_\alpha)$$

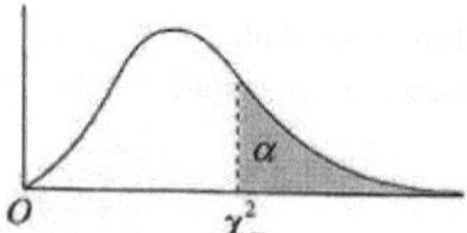

ν	$\alpha=0{,}990$	$\alpha=0{,}975$	$\alpha=0{,}950$	$\alpha=0{,}900$	$\alpha=0{,}100$	$\alpha=0{,}050$	$\alpha=0{,}025$	$\alpha=0{,}010$	$\alpha=0{,}001$
1	0,0002	0,0010	0,0039	0,0158	2,71	3,84	5,02	6,63	10,83
2	0,02	0,05	0,10	0,21	4,61	5,99	7,38	9,21	13,82
3	0,12	0,22	0,35	0,58	6,25	7,81	9,35	11,34	16,27
4	0,30	0,48	0,71	1,06	7,78	9,49	11,14	13,28	18,47
5	0,55	0,83	1,15	1,61	9,24	11,07	12,83	15,09	20,52
6	0,87	1,24	1,64	2,20	10,64	12,59	14,45	16,81	22,46
7	1,24	1,69	2,17	2,83	12,02	14,07	16,01	18,47	24,32
8	1,65	2,18	2,73	3,49	13,36	15,51	17,53	20,09	26,13
9	2,09	2,70	3,33	4,17	14,68	16,92	19,02	21,67	27,88
10	2,56	3,25	3,94	4,87	15,99	18,31	20,48	23,21	29,59
11	3,05	3,82	4,57	5,58	17,27	19,67	21,92	24,72	31,26
12	3,57	4,40	5,23	6,30	18,55	21,03	23,34	26,22	32,91
13	4,11	5,01	5,89	7,04	19,81	22,36	24,74	27,69	34,53
14	4,66	5,63	6,57	7,79	21,06	23,68	26,12	29,14	36,12
15	5,23	6,26	7,26	8,55	22,31	25,00	27,49	30,58	37,70
16	5,81	6,91	7,96	9,31	23,54	26,30	28,84	32,00	39,25
17	6,41	7,56	8,67	10,08	24,77	27,59	30,19	33,41	40,79
18	7,01	8,23	9,39	10,86	25,99	28,87	31,53	34,80	42,31
19	7,63	8,91	10,12	11,65	27,20	30,14	32,85	36,19	43,82
20	8,26	9,59	10,85	12,44	28,41	31,41	34,17	37,57	45,32
21	8,90	10,28	11,59	13,24	29,61	32,67	35,48	38,93	46,80
22	9,54	10,98	12,34	14,04	30,81	33,92	36,78	40,29	48,27
23	10,20	11,69	13,09	14,85	32,01	35,17	38,08	41,64	49,73
24	10,86	12,40	13,85	15,66	33,20	36,41	39,37	42,98	51,18
25	11,52	13,12	14,61	16,47	34,38	37,65	40,65	44,31	52,62
26	12,20	13,84	15,38	17,29	35,56	38,88	41,92	45,64	54,05
27	12,88	14,57	16,15	18,11	36,74	40,11	43,19	46,96	55,48
28	13,57	15,31	16,93	18,94	37,92	41,34	44,46	48,28	56,89
29	14,26	16,05	17,71	19,77	39,09	42,56	45,72	49,59	58,30
30	14,95	16,79	18,49	20,60	40,26	43,77	46,98	50,89	59,70

Quand ν est supérieur à 30, on utilise la table de la loi normale (table de l'écart réduit) avec :

$$t = \sqrt{2\chi^2} - \sqrt{2\nu - 1}$$

TD n° 04: Study of cohabitation

TD n° 4: Study of cohabitation

Introduction

For billions of years, the cohabitation of living beings (animals, plants, minerals, fungi and bacteria) on Earth has evolved into a complex, interconnected system, where each life form plays a vital role in the functioning of our planet and contributes to the overall balance and health of the Earth (Clobert et *al.*, 2012).

1. Definition

Coexistence defines the various biocenoses present in the surveyed area. The coexistence of various species in the same survey shows that the environment meets their common requirements. But it is only the repeated coexistence of these same species in numerous surveys that enables us to speak of a biocenosis.

Several parameters can be used to establish the degree of association between two species. The most commonly used are :

$$Q = \frac{2 * C * 100}{A + B}$$ **Sørensen index**

$$Q = \frac{C * 100}{A + B + C}$$ **Jaccard's index (or coefficient)**

$$..Q = \frac{A - B}{A + B}$$ **Odum index**

With :

A: number of surveys where species A is present ;

B: number of surveys where species B is present ;

C: number of surveys where both species are present.

Note

The coefficients only take into account the presence or absence of species. They do not don't take their abundance into account.

2. *The chi-square method (or test),χ^2*

Affinity between two species can be assessed using the chi-square method. First, we calculate P, which corresponds to the most probable number of samples in which two randomly grouped species can cohabit (Vaillant, 2005).

$$P = \frac{A * B}{N}$$

With :

A: number of surveys where species A is present ;

B: number of surveys where species B is present ;

N: total number of readings taken.

Next, P is compared with C (C= the number of surveys in which both species A and B are present).

There are three possible scenarios:

If P> C, there is exclusion;

If P= C, the two species are randomly distributed;

If P < C, the two species tend to cohabit with a certain probability. This is the assumption made beforehand.

In the latter case only (**P < C**), the hypothesis is tested using the chi-square test. To find out with what probability this cohabitation corresponds to something real and is not due to random sampling, we calculate the value of χ^2.

$$\chi^2 = N^3 (C - P)^2 / A * B (N - A)(N - B)$$

χ^2 allows you to compare a distribution arranged by classes with another distribution of identical arrangement. It is possible to compare :

- An observed distribution is a distribution calculated from a theoretical distribution;
- Two or more distributions between them.

The calculated chi-square (χ^2_c) is then compared with the chi-square from the distribution table (χ^2_t).

If $\chi^2_c < \chi^2_t$, the hypothesis is accepted (the two species cohabit);

If $\chi^2_c > \chi^2_t$, the hypothesis is rejected (the coexistence of the two species is due to the chance of sampling).

To read χ^2_t, use Pearson's distribution table, including the risk error of **5%** (0.05) and the corresponding **ddl** :

ddl = (number of classes of species A - 1) * (number of classes of species B - 1). ddl: Degree of

Freedom.

3. *Application exercise on cohabitation*

Table 4. Application exercise on cohabitation

Prel. / Esp.	1	2	3	4	5	6	7	8	9	10	11	12	13	14	15	16	17	18	19	20	21	22	23	24	25	26	27	28	29	30
A	6	1	5	3	0	0	7	6	3	2	0	1	2	3	12	0	8	10	0	18	17	10	0	9	0	0	14	17	0	9
B	18	7	2	1	0	3	4	0	5	5	0	7	10	6	10	8	0	9	14	15	17	1	0	19	0	18	17	1	2	0
C	0	0	0	0	0	3	5	0	0	9	0	1	0	0	0	8	2	7	0	0	16	15	0	0	19	17	13	0	0	0

Source: Lamine, 2013

Question: Study the cohabitation of the following pairs of species: A-B, A-C and B-C.

4. *Correction of the application*

P= a.b / N

$X^2= N^3$ / a.b (N-a) (N-b) x (C*- P).

a: number of samples where species A is present ;

b: number of samples where species B is present ;

C*: number of samples where species A and B are present;

P: most probable number of samples in which the two species (A and B) cohabit.

Species

P= a.b / N= 21.23 / 30= 16.1 C* = 18

P< C*. ➡ the two species tend to cohabit.

X^2 test.

- Hypothesis testing between A and B:

$X^2_{A.B}$= [30^3/ 21.23 (30-21) (30-23)]. $(18-16,1)^2$= 3,2

X^2_C= 3,2

X^2_t at 95% with ddl= (na-1) (nb-1)= (5-1). (5-1)= 4.4 =16

X^2_t= 26.3

$X^2_t > X^2_C$ ➡ Accepted hypothesis A and B live together.

Species A-C

P=a.b / N= 21.12 / 30= 16.1

C*= 8

P>C* ➡ exclusion.

P=a.b / N= 23.12 / 30= 9.2

C*= 10

P<C* ➡ both species tend to cohabit

Species B-C

X^2 test.

@ Hypothesis testing between A and B:

$X^2_{A.B}$= [30^3/ 23.12 (30-23) (30-12)]. $(10-9,2)^2$= 0,5

X^2_t at 95% with ddl= (na-1) (nb-1)= (5-1). (5-1)= 4.4= 16

$X^2_t > X^2_C$ ➡ Assumption accepted: the two species B and C cohabit.

Conclusion

The harmonious cohabitation of different life forms and the ecosystems to which they belong contributes to the balance and development of biodiversity. Biodiversity is the living fabric of the planet. It also refers to all the interactions that exist between different organisms, and between these organisms and their environment. Biodiversity therefore plays a fundamental role in the existence and evolution of Life on Earth.

TD n° 05: Biotic factors / intra- and interspecific relationships

TD n° 05: Biotic factors / intra- and interspecific relationships

Introduction

Competition occurs when organisms use common resources that are present in limited quantities or, if these resources are not limited, when competing organisms harm each other by seeking them out. As with intraspecific competition, competition between species for resources can be direct or indirect. We therefore speak of competition by interference and competition by exploitation.

1. Biotic interactions (Homotypic and heterotypic)

In ecology, biotic factors represent all the interactions between living organisms and living organisms in an ecosystem. Opposable to abiotic factors, they constitute part of the ecological factors of a given ecosystem. They include food resources, trophic relationships of predation, cooperation, competition, parasitism, etc. (Frontier et *al*., 2008).

We distinguish two categories of biotic factors that are determined by the types of relationships between living beings:

***1.1. Intraspecific (Homotypic) relationships*:** These are relationships between individuals within the same species.

***1.1.1. Group effect*:** Characterized by changes in animals of the same species that live in groups to defend themselves and gain access to environmental resources. It has been demonstrated in many insects (locusts) and vertebrates (birds and mammals).

***1.1.2. Mass effect*:** Characterized, in contrast to the group effect, by the fact that individuals of the same species get in each others way. This leads to overcrowding. In this case, individual aggressiveness increases, female fecundity decreases and collective suicidal actions (e.g. cannibalism) occur.

***1.1.3. Competition*:** Competition between individuals of the same species can take different forms. In animals, competition is established in the search food, shelter, reproduction sites, and more generally for the conquest and preservation of territory. Most territorial species restrict access to their territory to other species. Competition can be direct (intraspecific) or indirect (interspecific).

1.2. Interspecific relationships (Heterotypic)

1.2.1. Mutualism**:** The species can only live, grow or multiply in the presence of the mutual species. Benefits are shared by both species. Example: Association of plants with nitrogen-fixing bacteria, bacterial flora living in the digestive tract of animals).

1.2.2. Neutralism: Species that are independent of one another and have no influence on one another. E.g. Species with totally different ecological niches.

1.2.3. Symbiosis**:** Internal association between two partners. For example, mycorrhiza is a mutualism between a fungus and a plant. There is mutual benefit for both partners.

1.2.4. Predation**:** A predator will alter the abundance of prey species, but will not lead to prey to point of extinction, ensure its survival and the continuity of its resources.

1.2.5. Parasitism**:** Parasitism corresponds to a more or less close one-sided association with different organisms, with the parasite living at the expense of another living being called the host.

1.2.6. Amensalism**:** A species is inhibited in its growth or reproduction by another inhibiting species (amensal) which secretes toxic substances into the environment. This action is beneficial for the amensal species, but harmful for the host.

1.2.7. Commensalism**:** Association of a commensal species that benefits from another species that does not benefit or harm from it. E.g. Animals sharing the same habitat or using the food scraps of other animals.

1.2.8. Cooperation**:** Two or more species form a non-reciprocal cooperation, a beneficial relationship for the different cooperating species. Ex. Colonial birds.

2. Application exercise n° 01 on biotic factors

Place the following biotic relationships in the corresponding category in the table: Group effect - Parasitism - Amensalism - Interspecific competition - Cooperation - Mass effect -Commensalism - Mutualism.

Table 5. Application exercise on intra-specific relationships

Beneficial homotypic relationships	Harmful homotypic relationships	Beneficial heterotypic relationships	Harmful heterotypic relationships
....	....	....	....

Source: Dajoz, 2006

3. *Application exercise n° 02 on biotic factors*

Species coexist in a stand, with numerous interactions between them. If we consider two species A and B, several types of interactions can be determined, as shown in the table below:

Table 6. Application exercise on biotic factors

Interactions	Species A	Species B
Parasitism (A parasite B host)	....	....
Amensalism (B amensal A inhibited)	....	....
Competition (A and B: two competing species)	....	....
Neutralism (A and B: two neutral species)	....	....
Symbiosis (mandatory interaction)	....	....
Cooperation (interaction not mandatory)	....	....
Commensalism (A commensal, B host)	....	....
Predation (A predator, B prey)	....	....

Source: Dajoz, 2006

4. *Correction of exercises n° 01 & 02*

Exo 01 - Biotic factors

Beneficial homotypic relationships	Non-beneficial homotypic relationships	Beneficial heterotypic relationships	Beneficial non-heterotypical relationships
Group effect	Mass effect	Cooperation Mutualism	Parasitism Amensalism Interspecific competition Commensalism

Exo 02 - Biotic factors

Interactions	Species A	Species B
Parasitism (A parasite B host)	+	-
Amensalism (B amensal A inhibited)	-	**0**
Competition (A and B: two competing species)	-	-
Neutralism (A and B: two neutral species)	**0**	**0**
Symbiosis (mandatory interaction)	+	+
Cooperation (interaction not mandatory)	+	+
Commensalism (A commensal, B host)	+	**0**
Predation (A predator, B prey)	+	-

0: Species unaffected;**+ :** Species life improved; **-:** Species life disrupted.

Conclusion

In a stable environment, competition between species using the same resources can lead to the elimination of one species by another. This is the principle competitive exclusion, also known as the Gause principle. Coexistence is only possible when ecological niches are sufficiently distinct.

TD no. 06: Food chains and food webs

TD no. 06: Food chains and food webs

Introduction

In an ecosystem, the structure of food webs (the types and networks of food relationships between organisms) strongly influences the quantity, diversity, stability and quality of biomass and residual organic matter (excreta, necromass) produced by ecosystems (Canuel *et al.*, 2007). The quality of a food web and its interactions is directly linked to the stability and resilience of the populations it supports (Ives & Cardinal, 2004).

1. Trophic relationships between living beings

Within ecosystems, living beings ensure their survival establishing relationships with each other. complex trophic systems.

Table 7. Some examples of relationships between living beings.

Food relations	Examples
Aphids are small insects that depend on plants to suck their sap, causing considerable damage to plants. plants.	
The ladybug is a small crab that feeds aphids.	

The Pinnothère is a small crab that protects itself in the mussel's shell and eats among the plankton filtered by the mussel. The presence of this crab does not harm the mussel.

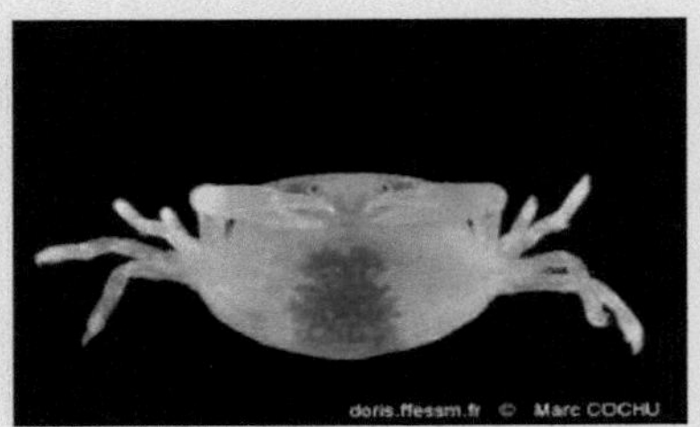

In nature, ants are often seen around a group of plant aphids: the ants like the aphids, which provide a sugar-rich honeydew, and at the same time the ants protect the aphids.

Termites feed on wood cellulose. If the termites are exposed to a high concentration of oxygen, which kills the protozoa but leaves the termite alive, the latter dies a few days later from starvation.

Growing oilseed radish and wheat in the same environment reduces growth yields by 36% for radish.
But grown separately, their yields increase.
Radishes, like wheat, are nitrogen-hungry.

Many fungi develop and grow on trees or dead leaves.

Source: https://www.encyclopedie-environnement.org/vivant/systemes-symbiotiques-parasites/.

2. *Food chains*

According to Emmerson et *al* (2005), a food chain is a sequence of living organisms at different trophic levels, each of which eats organisms at a lower trophic level in order to acquire energy. The first link in a chain is always an autotrophic organism. In seas and oceans, phytoplankton plays this role. In the abyssal depths, where the sun's rays cannot reach, thermophilic bacteria are the first links in the chain. However, the photosynthetic chain still exists there, as pelagic organisms die and sink. Humans are often the last link in the chain: they are super predators.

3. *Food webs*

A food web is a set of interconnected food chains within an ecosystem through which energy and biomass circulate (exchange elements such carbon and nitrogen flows between the different levels of the food chain, exchange of carbon between autotrophic and heterotrophic plants) (Petchy et *al*., 1999). Food webs are affected by global changes, including those linked to climate disruption, at the level of keystone species (Sanford, 1999).

4. *Exercise n° 01*

From the following network :

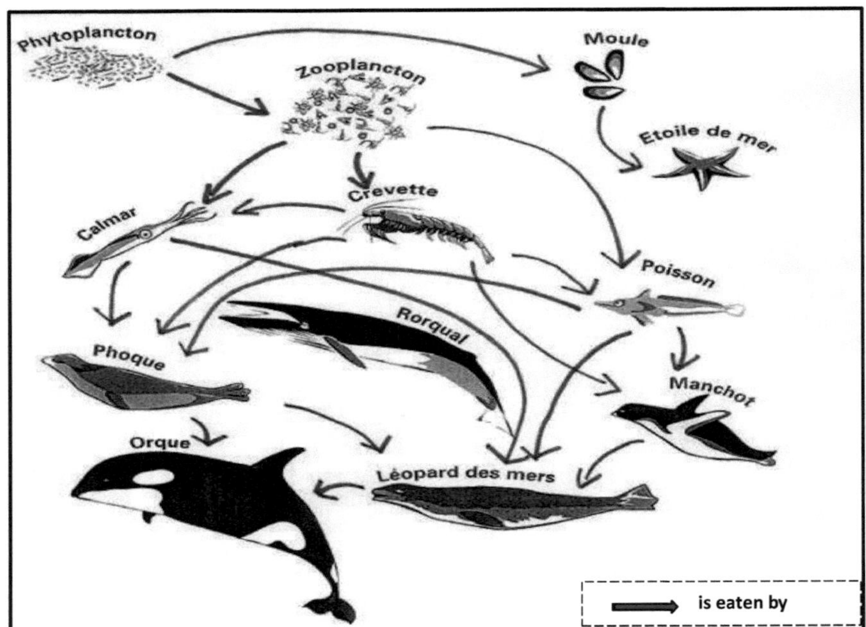

Figure 5. Dynamics of a marine food web

a) Identify the ultimate carnivore

b) Identify the species that eat shrimp.

c) Identifies the producer(s)

d) What would happen to other organisms if the seal population were to disappear due to uncontrolled hunting?

e) From the network above, draw a food chain.

f) How many trophic levels are there in the previous food chain?

g) Draw an energy pyramid of the food chain (e), knowing that there are 50,000 kJ of energy in phytoplankton.

5. *Exercise n° 02*

Using the food chain below, answer following questions.

Cereals□ Souri□ Snake□ Fox□ Wolf

1) To which trophic level does the first organism belong?

2) Which of these organizations is the secondary consumer?

3) Which one is the ultimate consumer?

6. Exercise n° 03

Match each statement in column A with the correct term in column B by writing the letter in the space provided.

1. An organism belonging to the highest trophic level in a food chain
2. A linear sequence interactions between organisms
3. Food level
4. The substance that growers capture from the Sun
5. An organism that feeds on plants and animals
6. An organism capable of producing its own food
7. An organism that only eats meat
8. An organism that is hunted and eaten
9. A network of interrelated food chains
10. Organisms that recycle valuable nutrients
11. An organism that depends on other organisms for food
12. An organism that eats only plants

a) a carnivore of the last order

b) a food chain

c) a carnivore

d) an omnivore

e) trophic level

f) a decomposer

g) an ecosystem

h) biomass

i) energy

j) a prey

k) the consumer

l) a food network

m) a producer

n) a herbivore.

7. *Exercise n° 04*

Fill in the small boxes and give the different trophic levels of this ecosystem.

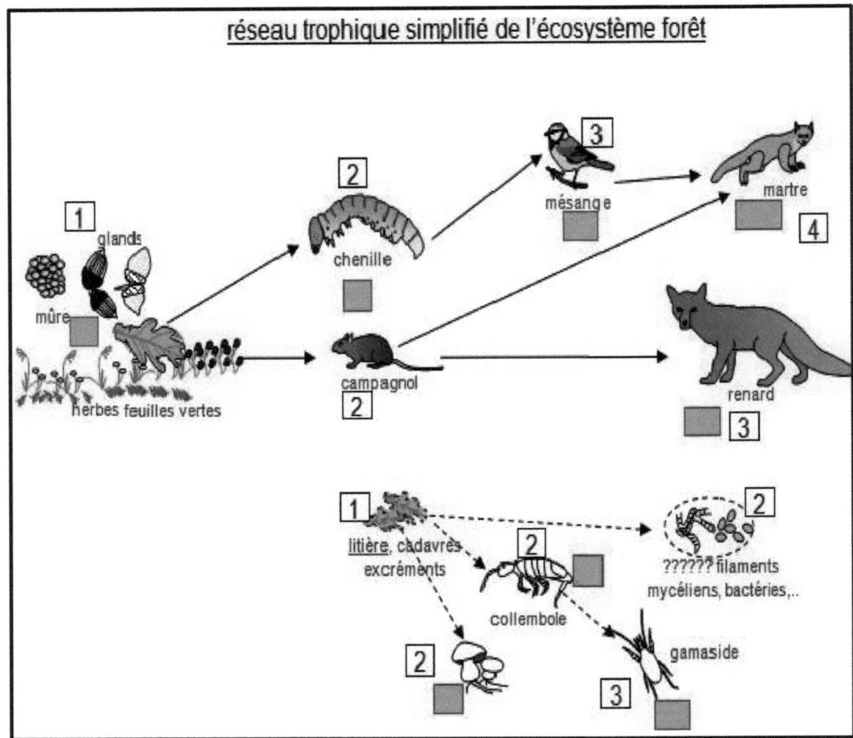

Figure 6. Simplified trophic network of a forest ecosystem

8. *Correction of exercises 01, 02, 03 and 04.*

- **Correction - exercise n° 01**

a) Identifies the ultimate carnivore: **Orca**__________________________

b) Identify the species that eat shrimp. Fish**, Penguin, Seal, and Squid**

c) Identify the producer(s): **Phytoplankton**

d) What would happen to other organisms if the seal population were to disappear due to uncontrolled hunting: **Decrease in : Zooplankton, killer whales and leopard seals.**

e) From the network above, draw a food chain.

Phytoplankton Zooplankton · · Shrimp · Seal · Whale

f) How many trophic levels are there in the previous food chain?

5 trophic levels.

g) Draw an energy pyramid of the food chain (e), knowing there are 50,000 kJ of energy in the food chain.

in phytoplankton

Orca (5 kJ)
Seal (50 kJ)

Shrimp (500 kJ)
Zooplankton (5,000 kJ)
Phytoplankton (50,000 kJ)

- **Correction - exercise n° 02**

Using the food chain below, answer the following questions.

Cereals Souri · · Snake · Fox · Wolf.

1) To which trophic level does the first organism belong? **Producers**
2) Which of these is the secondary consumer? **Garter snake**
3) Which one is the ultimate consumer? **Loup**

- **Correction - exercise n° 03**

Match each statement in column **A** with the correct term in column **B** by entering the letter in the space provided.

________A__ 1. An organism belonging to the highest trophic level in a food chain.

________B__ 2. A linear sequence interactions between organisms

________E__ 3. Food level

________I . The substance that growers capture from the Sun

___D__ 5. An organism that feeds on plants and animals

___M__ 6. An organism capable of producing its own food

___C__ 7. An organism that only eats meat

___J__ 8. An organism that is hunted and eaten

___L__ 9. A network of interrelated food chains

___F__ 10. Organisms that recycle valuable nutrients

___K__ 11. An organism that depends on other organisms for food

___N__ 12. An organism that only eats plants.

- **Correction - exercise n° 04**

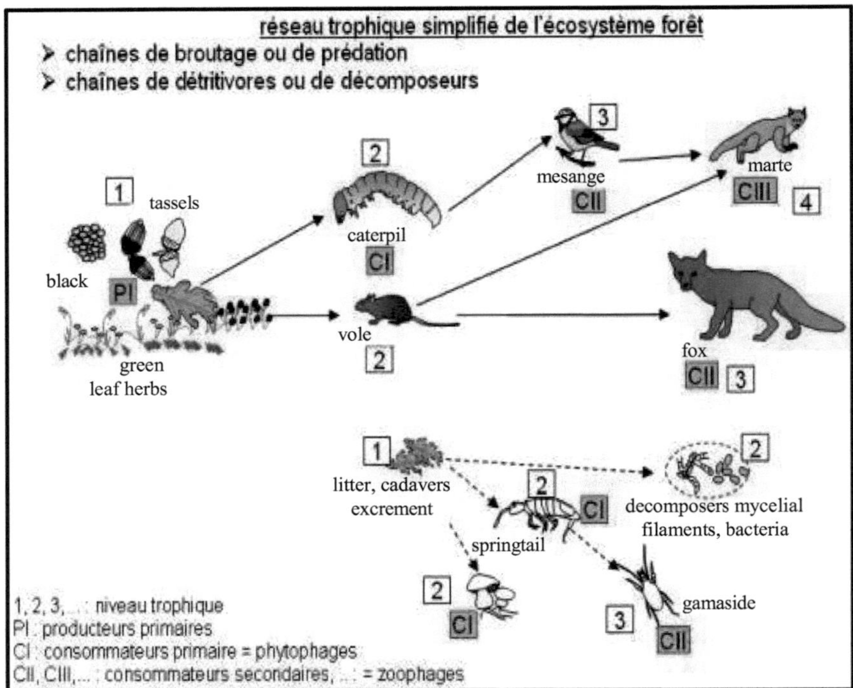

1,2,3....Trophic **levels**
PI: Primary producers
CI : Primary consumers= Phytophages
CII, CIII,. Secondary consumers,= zoophagous.

Conclusion

A food web, when viewed from the angle of biotic interactions, is the set of species united by trophic links, coexisting in a given place, as well the trophic links that unite them. Food webs are therefore the carriers of energy and matter flows that traverse communities.

TD n° 07: Geographical distribution of certain rodent species according to climatic conditions

Part 7 Geographical distribution of certain rodent species according to climatic conditions

TD N° 07: Geographical distribution of certain rodent species according to climatic conditions

Introduction

Biogeography is the study of the geographical distribution of species. This field of ecology determines the potential and actual distribution areas of different species, as well as how these areas evolve as a function of variations in the environment or human activities. Biogeography is also linked to the history of the Earth and the evolution its crust and biosphere. Indeed, a species' range is not only explained by its biological characteristics, but also by the evolution of its geography and evolutionary history (De Broyer, 2014).

1. ***TD aim***: Determination or delimitation of the distribution air of 3 species of Ctenodactylidae from climatic data.

2. ***Choice of species***: Three species of Ctenodactylidae were chosen for this study:

- *Ctenodactylus gundi*
- *Ctenodactylus vali*
- *Ctenodactylus m'zabi*

The genus *Ctenodactylus* has two species in Algeria, *Ctenodactylus gundi* and *Ctenodactylus vali*. The *Massoutiera* genus has just one species: *Massouteria m'zabi*.

It should be noted that each of these two species differs from other in two respects

1. Tympanic bubble size;
2. Teething.

Gundis are rodents that live in relatively warm regions, similar in size to small rats, but with rabbit-like hair and tails. These species of Ctenodactylidae are very interesting from a physiological point of view, as they don't seem to drink water (yet they live in the South).

3. ***Study***

The study method is trapping. In general, for this type of mammal, swing cages are used. This method allows the capture of several individuals, giving the most accurate estimate for these populations.

As for range, this depends on the home range of the species under study. Gundis live in a nest or burrow, which we always have to look for. They never dig holes, often taking shelter under rocky screes, and can also cohabit with man.

4. *Emberger*

For these Ctenodactylidae, territoriality is unknown. We interested in placing the different species of Ctenodactylidae (*Ctenodactylus gundi, Ctenodactylus vali* and Ctenodactylus m'zabi) on the Emberger Climagram according to the most important climatic data (rainfall and temperature).

5. *Rainfall quotient (Emberger)*

$$Q2 = \frac{1000P}{(\frac{M+m}{2})(M-m)}$$

P: annual rainfall in mm ;

M: maximum annual temperature in degrees Celsius; **m**: minimum annual temperature in degrees Celsius; **M + m /2** = average annual temperature; **M-m** = thermal amplitude.

The greater the annual dryness, the lower the Quotient value.

Note

Emberger formula (Mediterranean climate) has been simplified by Stewar for Algeria.

$$Q2 = 3{,}43\frac{p}{M-m}$$

6. *Application exercise*: Geographical distribution of rodent species according to climatic data.

Data :

Table 8. Application exercises on the geographical distribution of certain species according to climatic data.

Stations / Index	1	2	3	4	5	6	7	8	9	10	11	12	13	14	15	16	17	18	19	20
Q2	29	6	27	8,6	5,6	9,5	12	29	2,1	16	17	25	13	3,2	5,2	19	24	5,6	38	7,1

m	4,5	4,6	5,3	4,9	5,7	6	6	6	6,2	6,4	7,1	3	3,4	3,5	3,6	3,6	3,7	4,3	0,8	1,8
Species surveyed	○	■	○	○	●	●	○	○	■	○	●	○	●	●	■	○	○	■	○	●

Source: Meddour et *al.*, 2019

Captions:

Cten○dactylus gundi
Cten■dactylus m'zabi

Cten●dactylus vali

7. *Correction of the application*

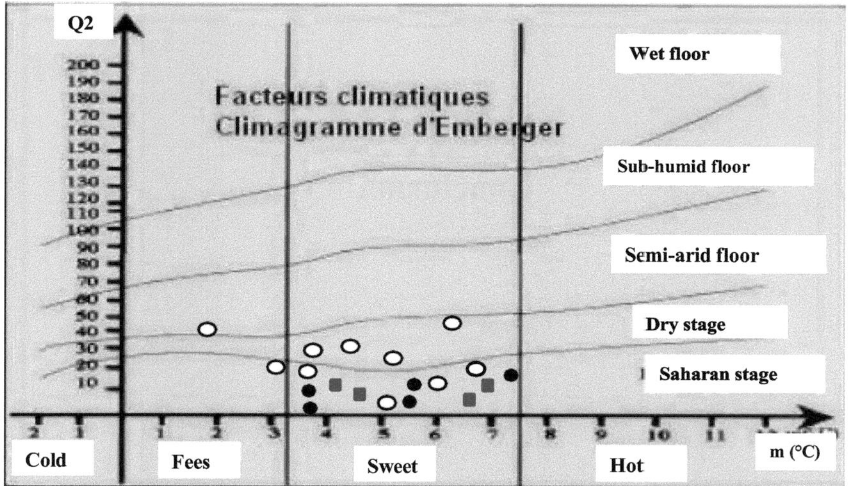

Figure 7. Emberger

Conclusion

The current distribution of species and ecosystems is always more or less a reflection of past distributions. For example, recent fluctuations in Quaternary climate and the spatial dynamics of biodiversity they engendered are still evident in the structure and dynamics of today's ecosystems.

Conclusion

General conclusion

We have seen that living beings are subject to various abiotic and biotic factors in their environment. They are also in constant interaction with their biotope, creating ecological balances between them. Man, as a part of nature, despite all the technologies he has developed, cannot escape and consider himself the master of all living beings. He cannot escape the natural laws that govern the lives of the animal and plant species living on planet earth. There is only one earth, and we must protect it.

"Ecology is also, and above all, a cultural problem. Respect for the environment requires a large number of changes behaviors.

Nicolas Hulot

References

References

- **Barbault R. 2008**. Ecologie générale. Structure et fonctionnement de la Biosphère. 6th edition DUNOD.

- **Canuel E.A., Spivak A.C., Waterson E.J. & Duffy J.E. 2007**. Biodiversity and food web structure influence short-term accumulation of sediment organic matter in an experimental seagrass system; *Limnol. Ocean.* **52**, 590- 602.

- **Clobert J., Baguette M., Benton T.G. & Bullock J.M. 2012**. Dispersal ecology and evolution. Oxford University Press.

- **Dajoz R. 2006**. Précis d'Ecologie. 8th edition. Edition Dunod. 631p.

- **De Broyer C. 2014**. Biogeographic Atlas of the Southern Ocean. Scientific Committee on Antarctic Research.

- **Emmerson M., Bezemer M., Hunter Md. & Jones TH. 2005.** Global change alters the stability of food webs. *Global Change Biology*, **11**, 490-501.

- **Faurie C., Ferra C., Medori P., Devaux J. & Hemptinne J.L. 2011.** Ecologie : Approche scientifique et pratique. 5ème édition. Edition Lavoisier. 407p.

- **Frontier S., Pichod-Viale D., Lepretre A., Davoult D. & Luczak C. 2008.** Ecosystème : Structure, Fonctionnement, Evolution. 4th edition. Edition Dunod. 558p.

- **Ives A. & Cardinale B. 2004.** Food-web interactions govern the resistance of communities after non-random extinctions. *Nature* (review), **429**, 174-177.

- **Lamine S. 2013**. Contribution à la connaissance des macro-invertébrés benthiques du sous-bassin versant des Ouadhias. Master's thesis, Université Mouloud Mammeri de Tizi- Ouzou, 67pp.

- **Lamine S. 2021.** Research on the faunistics, ecology and biogeography of Ephemeroptera, Plecoptera, Trichoptera and Coleoptera Elmidae & Hydraenidae. Doctoral thesis 3th cycle, Université Mouloud Mammeri de Tizi-Ouzou, 288 pp.

- **Marc Mazerolle J. 2019**. Introduction to sampling theory, Web.

- **Meddour S., Mlik R. & Sekour M. 2019**. Characterization of the abundance of Atlas Goundi *Ctenodactylus gundi* (Rodentia, Ctenodactylidae) in the southern Aurès (eastern Algeria). *International Journal of Natural Resources and Environment*. Vol. **1**, No. 1; pp. 29-35.

- **Petchey Ol., Mcphearson Pt., Casey Tm. & Morin PJ. 1999.** Environmental warming alters food-web structure and ecosystem function. *Nature*, 402, 69-72.

- **Ramade F. 2009.** Element of Ecology: Fundamental Ecology. 4th edition. Edition Dunod. 689p.

- **Sanford E. 1999.** Regulation of keystone predation by small changes in ocean temperature.
Science, **283**, 2095- 2097.

- **Tirard C., Abbadie L., Laloi D. & Koubbi PH. 2016**. Ecologie, Licence, Master & CAPES. DUNOD, 11, rue Paul Bert, 92240 Malakoff. www.dunod.com.

- **Tirard C., Barbault R., Abaddie L. & Loeuille N. 2012**. Mini manual d'écologie. DUNOD.

- **Vaillant J. 2005**. Initiation à la théorie de l'échantillonnage, web.

MIX
Papier aus verantwortungsvollen Quellen
Paper from responsible sources
FSC® C105338

Printed by Books on Demand GmbH, Norderstedt / Germany